Slide, flip, rotate, turn!
With some special shapes you'll learn
how to pattern, how to tile—
how to arrange shapes with style!

Let's meet some shapes—
some old, some new.
Then you'll see
what they can do!

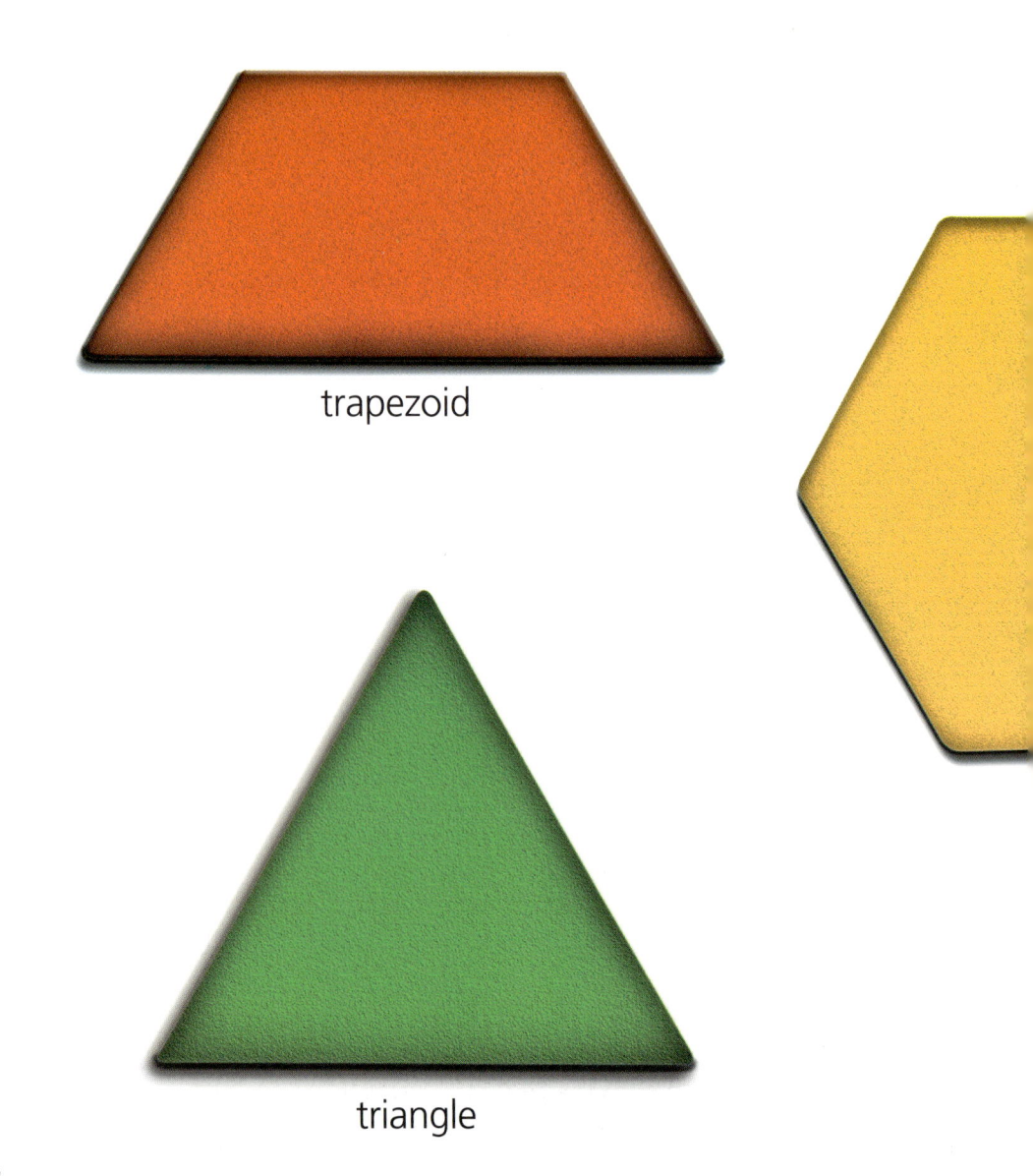

trapezoid

triangle

Find the **trapezoid**, **hexagon**, **triangle**, and **square**.
And don't forget the **rhombus**—
it's also there.

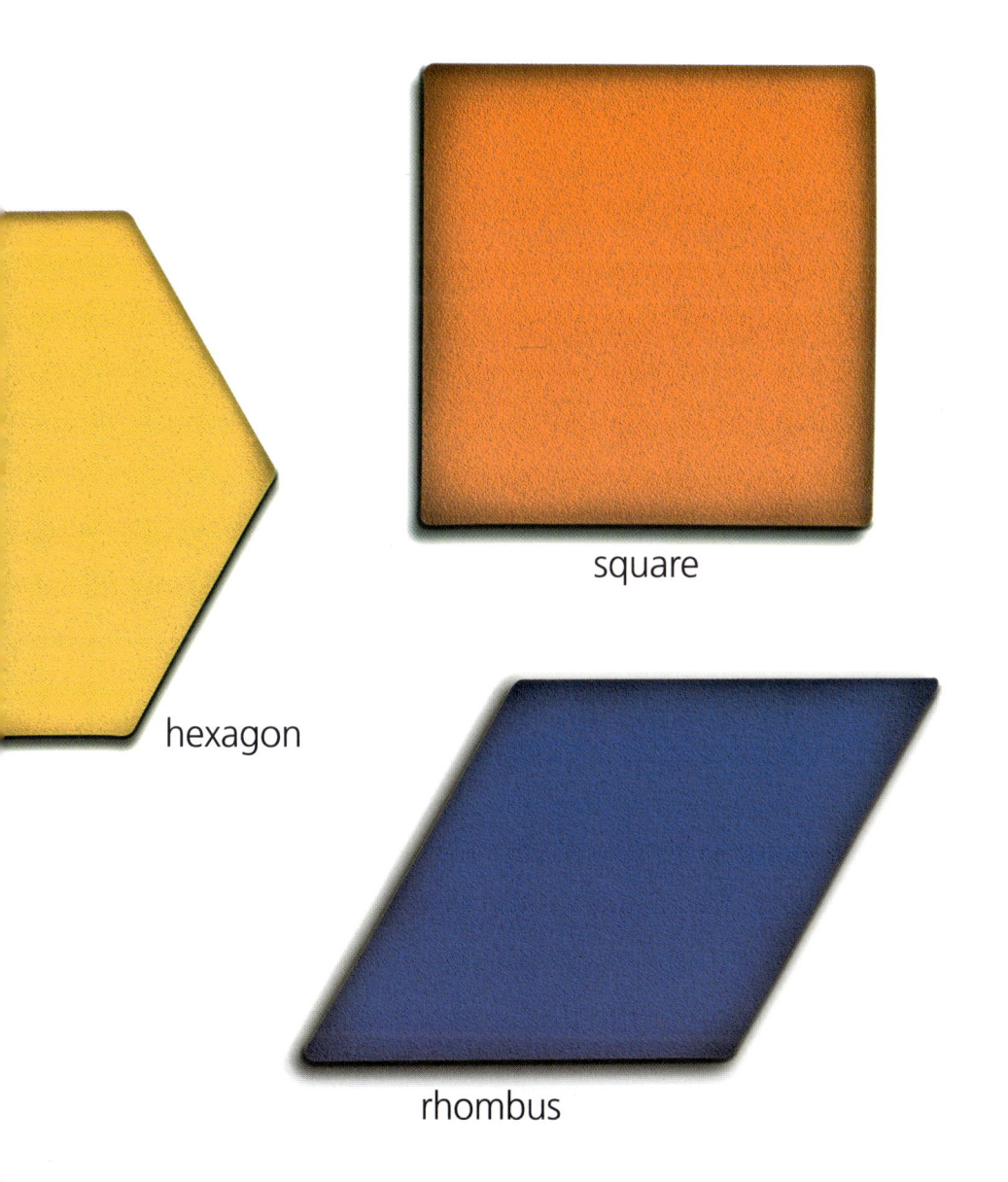

square

hexagon

rhombus

Turn shapes to the left or right, flip them upside down.

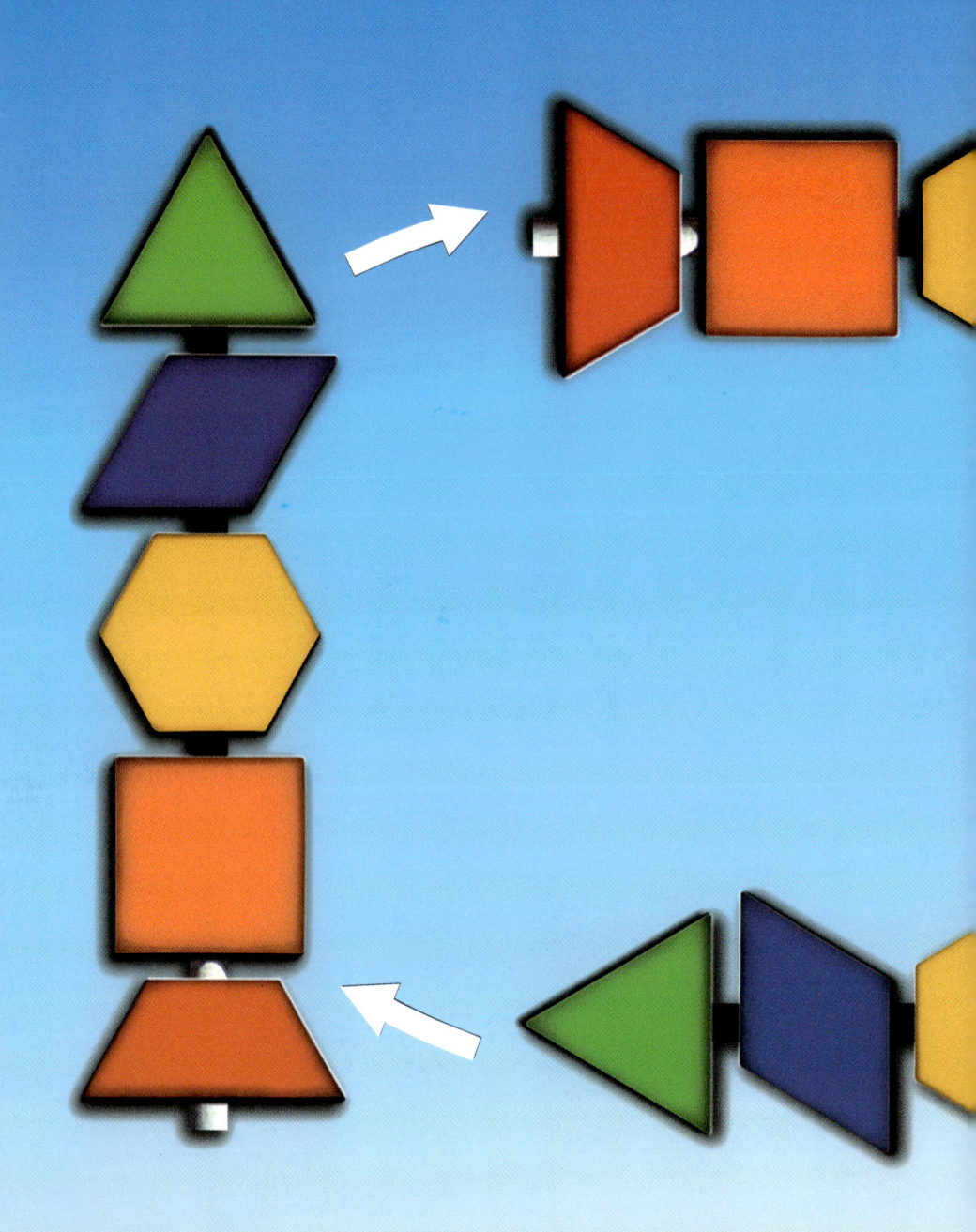

What happens when you rotate a shape?
Watch as they turn around.

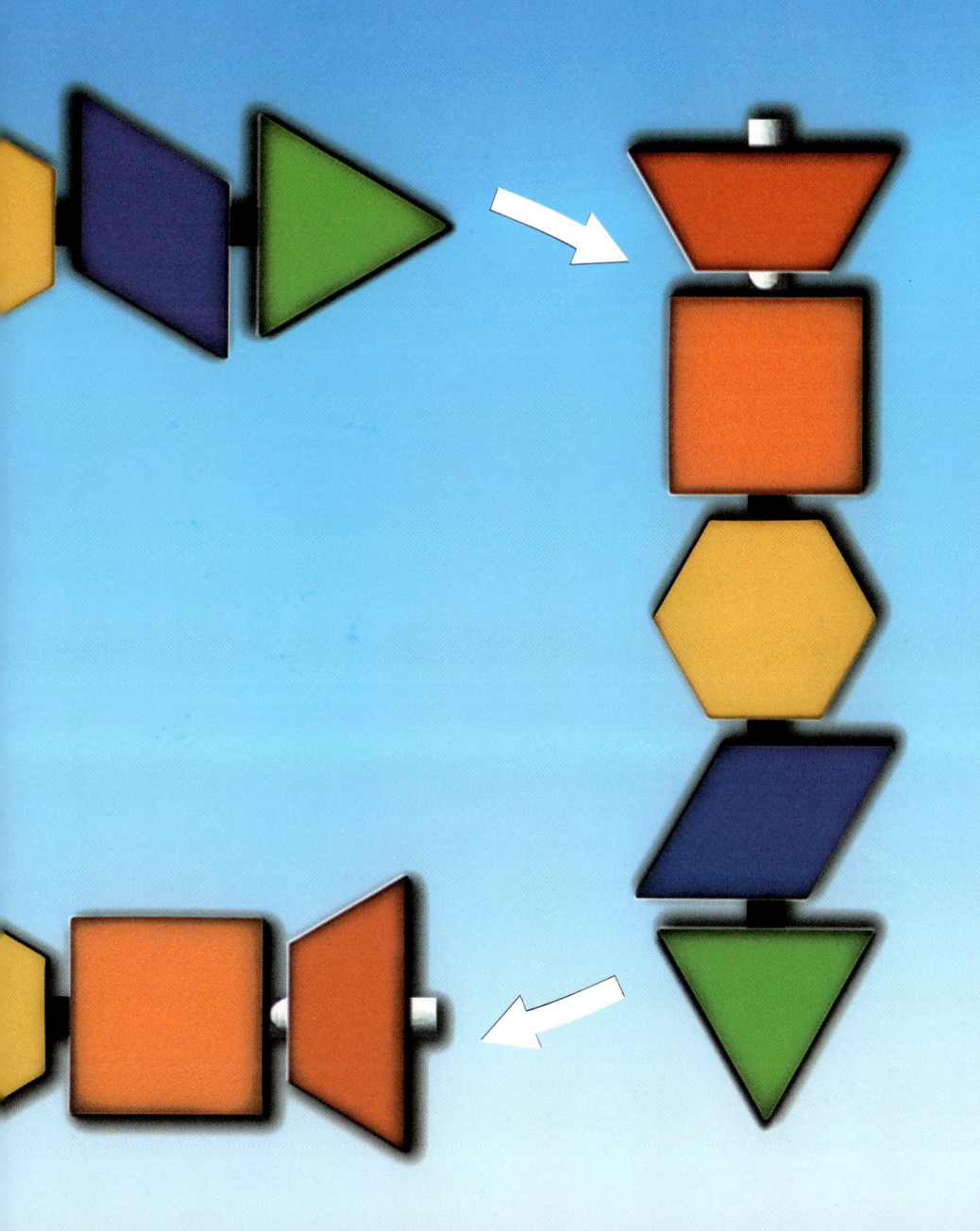

5

You can pattern, or repeat, these shapes in an orderly way.

From left to right or top to bottom, which shapes repeat here? Can you say?

When shapes fit together
with no space in between,

they form a **tiling** pattern.
Can you see what tiling means?

Does tiling happen naturally?
Believe it or not, it can!
You know that bees tile with hexagons,
if you're a geometry fan.

Now can you name the other shapes that make up a honeycomb? It's easy to see the busy bee has a remarkable home!

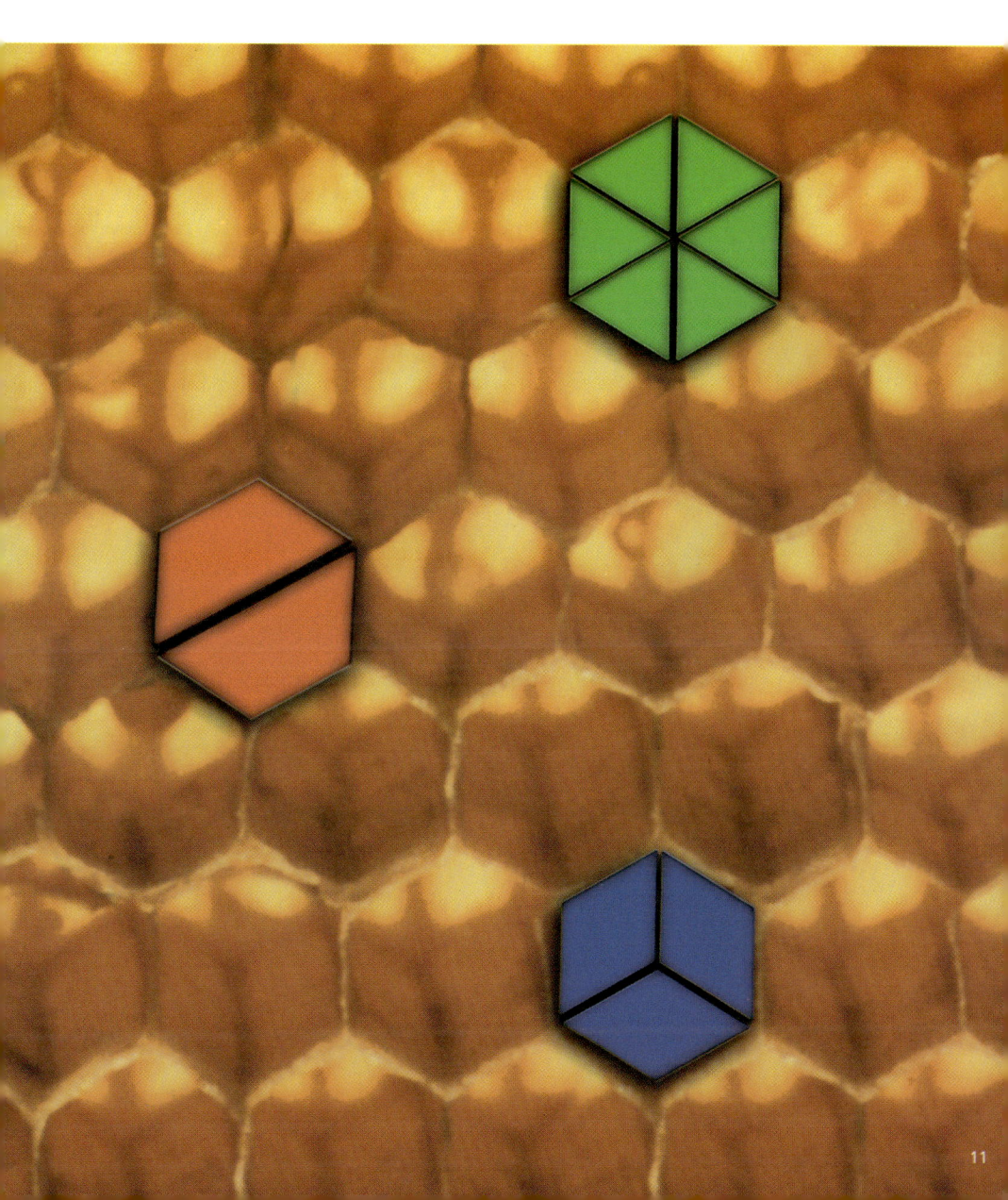

When a pattern or shape is *balanced* like this mosaic that you see,

from left to right or top to bottom—
it is called **symmetry**.

Which patterns are made by nature, which by woman or man?

What shapes are used in all these patterns?
Find the symmetry if you can!

Slide, flip, rotate, turn!
Now you have begun to learn
how to pattern, how to tile—
how to arrange shapes with style!